Yuri Kornyushin

Studying Quantum Mechanics Selected Topics

AF135945

Yuri Kornyushin

Studying Quantum Mechanics
Selected Topics

LAP LAMBERT Academic Publishing

Impressum / Imprint

Bibliografische Information der Deutschen Nationalbibliothek: Die Deutsche Nationalbibliothek verzeichnet diese Publikation in der Deutschen Nationalbibliografie; detaillierte bibliografische Daten sind im Internet über http://dnb.d-nb.de abrufbar.

Alle in diesem Buch genannten Marken und Produktnamen unterliegen warenzeichen-, marken- oder patentrechtlichem Schutz bzw. sind Warenzeichen oder eingetragene Warenzeichen der jeweiligen Inhaber. Die Wiedergabe von Marken, Produktnamen, Gebrauchsnamen, Handelsnamen, Warenbezeichnungen u.s.w. in diesem Werk berechtigt auch ohne besondere Kennzeichnung nicht zu der Annahme, dass solche Namen im Sinne der Warenzeichen- und Markenschutzgesetzgebung als frei zu betrachten wären und daher von jedermann benutzt werden dürften.

Bibliographic information published by the Deutsche Nationalbibliothek: The Deutsche Nationalbibliothek lists this publication in the Deutsche Nationalbibliografie; detailed bibliographic data are available in the Internet at http://dnb.d-nb.de.

Any brand names and product names mentioned in this book are subject to trademark, brand or patent protection and are trademarks or registered trademarks of their respective holders. The use of brand names, product names, common names, trade names, product descriptions etc. even without a particular marking in this work is in no way to be construed to mean that such names may be regarded as unrestricted in respect of trademark and brand protection legislation and could thus be used by anyone.

Coverbild / Cover image: www.ingimage.com

Verlag / Publisher:
LAP LAMBERT Academic Publishing
ist ein Imprint der / is a trademark of
OmniScriptum GmbH & Co. KG
Heinrich-Böcking-Str. 6-8, 66121 Saarbrücken, Deutschland / Germany
Email: info@lap-publishing.com

Herstellung: siehe letzte Seite /
Printed at: see last page
ISBN: 978-3-659-68139-4

Contents

Introduction

We start with analysing a concept of kinetic energy in Quantum Mechanics. Kinetic energy is a non-zero positive value in many cases of bound states, when a wave function is a real-valued one and there are no visible motion and flux. This can be understood, using expansion of the wave function into Fourier integral, that is, on the basis of virtual plane waves.

Then the ground state energy of a hydrogen atom is calculated in a special way, regarding explicitly all the terms of electrostatic and kinetic energies. The correct values of the ground state energy and the radius of decay are achieved only when the electrostatic energies of the electron and the proton (self-energies) are not taken into account. This proves again that self-action should be excluded in Quantum Mechanics.

Then a model of a spherical ball with uniformly distributed charge of particles is considered. It is shown that for a neutral ball (with compensated electric charge) the electrostatic energy is a non-zero negative value in this model. This occurs because the self-energy of the constituting particles should be subtracted. So it is shown that the energy of the electric field does not have to be of a positive value in any imaginable problem.

Then the problem of low-dimensional movements in Quantum Mechanics is considered.

Fermi and kinetic energy are usually calculated in periodic boundary conditions model, which is not self-consistent for low-dimensional problems, when particles are confined. Thus for confined particles the potential box model was used here self-consistently to calculate Fermi and kinetic energies in 3-, 2-, and 1-dimensional cases. This approach is much more logical and self-consistent.

Then the criteria of neglecting dimensions, that is the criteria of the 2- and 1- dimensional quantum movements, were derived.

The temperature criterion of low-dimensional quantum movements has been derived also.

Then rotation of the nucleus and rotation of the electronic cloud of the atom/ion were considered. It was shown that these rotations are not practically possible.

Then rotation of the cloud of delocalized electrons and ionic core of a fullerene molecule and these of the ring of a nanotube were discussed. It was shown that the rotation of the cloud of delocalized electrons of a fullerene molecule is possible and it goes in a quantum way when temperature is essentially lower than 40 K.

From the other hand rotation of the ion core of a fullerene molecule is possible in a classical way only. The same should be said about rotations in the ring of a nanotube.

Then a simple model for the autolocalization of a free charged particle is presented. The polarization well in the model is deep enough for only one localized level. In dielectric materials with a sufficiently large dielectric constant, two charged identical particles can be localized in one polarization potential well, forming a bipolaron.

Although several localized levels can be found in more realistic self-consistent models of this type, the more realistic theories require a high level of knowledge of mathematics. Hence, the proposed model can serve as an introduction to the ideas and concepts of autolocalized states.

In the end of the book a simple explanation of coexistence of zero angular momentum and non-zero magnetic moment in many-electron system is discussed from the point of view of Statistical Physics and Quantum Mechanics.

1. Kinetic and Electrostatic Energies in Quantum Mechanics

Introduction

Let us write down Schrödinger equation for an elementary particle [1]:

$$-(\hbar^2/2m)\Delta\psi + U(\mathbf{r})\psi = E\psi, \qquad (1\text{-}1)$$

where $\hbar$ is Planck constant, divided by 2π, m is the mass of a particle, Δ is Laplace operator, ψ is a wave function, $U(\mathbf{r})$ is the potential energy of a particle (in external with respect to the regarded particle itself field) as a function of a radius-vector $\mathbf{r}$, and E is the energy of the stationary state of a particle. In a case of a free particle $U(\mathbf{r}) \equiv 0$ and Eq. (1-1) has a well-known solution, a plane wave [1], $\psi(\mathbf{r}) = \text{const}(\exp i(\mathbf{k},\mathbf{r}))$, where i is the imaginary unity, and $\mathbf{k}$ is the wave-vector. This solution corresponds to a non-localized free particle. The energy of this particle is well known to be $E(k) = \hbar^2 k^2/2m$ [1]. This is a kinetic energy of a non-localized particle [1].

Potential energy $U(\mathbf{r})$ in Schrödinger equation, Eq. (1-1), does not include the potential energy of the interaction of the considered elementary particle with the field, produced by this same particle. Thus, in Quantum Mechanics the action of the elementary particle on itself, that is a self-action, is not taken into account. Equation (1-1) yields that $E = \langle T \rangle + \langle U(\mathbf{r}) \rangle$, where $T = -(\hbar^2/2m)\Delta$ is the kinetic energy operator, and $\langle...\rangle$ denotes quantum mechanical averaging [1]. We see that the stationary state energy of a particle does not include the potential energy of a particle itself.

Let us consider a charged particle, which produces electric field. The energy of this field is not included in the energy of the stationary state of a particle. This is because the energy of the electric field, produced by the particle, is not present in Schrödinger equation. It should be noted that this refers only to the particles charged with elementary charge $\pm e$, such as electron, proton, positron, antiproton, etc. When regarded particle consists of, for example, two protons, the field, produced by each proton, is an external one with respect to the other proton, and the potential energy of the interaction of the two protons should be introduced in Schrödinger equation. Thus, the charged elementary particle does not possess electrostatic energy in Quantum Mechanics.

When a charged particle is localized, it has a kinetic energy and creates an electric field. The energy stored in this field should not be taken into account when the total energy of a system is calculated, as taking it into account would be taking a self-action into account. So here we have electric field with no stored energy.

In the present section we shall discuss the kinetic energy when there is no flux/current, the electrostatic field without any electrostatic energy stored in it, and electrostatic energy when there is no electrostatic field. It shows again and again how wonderful Quantum Mechanics is.

Kinetic Energy in a General Case

Kinetic energy is the energy of a motion [1]. A matter flux and electric current accompany the motion of a charged matter. Matter flux $\mathbf{J}$ is given by the following equation [1]:

$$\mathbf{J} = (i\hbar/2m)(\psi\nabla\psi^* - \psi^*\nabla\psi). \tag{1-2}$$

It seems that no motion without a flux of a matter exists. In many of the bound states (both ground states and excited ones) of any particle, e.g., electron, the wave functions are real-valued, they contain no imaginary parts [1]. As it follows from Eq. (1-2), the real-valued wave function yields no flux and no electric current for a charged particle. As an example let us take an electron in the ground state of a hydrogen atom. Corresponding wave function is [1] $\psi(r) = (g^3/\pi)^{1/2}\exp{-gr}$, where $g = me^2/\hbar^2$ and r is the distance from the proton [1].

The operator of the momentum is $-i\hbar\nabla$, where ∇ is the gradient operator. The kinetic energy operator T is the operator of the momentum in square, divided by $2m$ [1]. This operator, $-(\hbar^2/2m)\Delta$, averaged quantum-mechanically over the whole space, is an average kinetic energy, $\langle T \rangle$ [1]. From this follows that for the electron in the ground state of a hydrogen atom $\langle T \rangle = \langle T \rangle_0 = me^4/2\hbar^2$. So we have here a very essential kinetic energy with no visible motion. This can be explained using virtual plane waves expansion.

As was mentioned above, the operator $-(\hbar^2/2m)\Delta$, averaged over the whole space yields kinetic energy. This energy is not zero because the wave function is inhomogeneous. It depends on coordinates. Let us expand the real-valued wave function into Fourier integral:

$$\psi(r) = \int a_k[\exp i(\mathbf{k},\mathbf{r})]d\mathbf{k}. \qquad (1\text{-}3)$$

The expansion coefficients are normalized, so that

$$\int a^*_k a_k d\mathbf{k} = 1. \qquad (1\text{-}4)$$

The kinetic energy operator $-(\hbar^2/2m)\Delta$, averaged over the whole space, is the kinetic energy of a system. Using Eq. (1-3) we get:

$$\langle T \rangle = (\hbar^2/2m)\int k^2 a^*_k a_k d\mathbf{k}. \qquad (1\text{-}5)$$

Equation (1-5) shows that the kinetic energy is a sum of kinetic energies of constituting plane waves with the appropriate weight.

Hydrogen Atom

Let us consider in this section a simple problem of a free hydrogen atom in a ground state. Let us do it in an alternative way, like it was done in [2]. Let us write the wave function of the electron in the ground state of a hydrogen atom in a form [2]:

$$\psi(r) = [\exp(-r/2R)]/2(2\pi)^{1/2}R^{3/2}, \qquad (1\text{-}6)$$

where R is related to the equilibrium (or not) radius of decay.

The energy of the ground state of the electron in a hydrogen atom is described by a well-known relation [1]:

$$E_{0h} = -(me^4/2\hbar^2), \qquad (1\text{-}7)$$

where e is the electron (negative) charge and m is the electron mass.

The electrostatic field created by the electron shell with the electric charge density

$$\rho(r) = e\psi^*(r)\psi(r) = (e/8\pi R^3)\exp(-r/R), \qquad (1\text{-}8)$$

can be calculated using Gauss theorem. The corresponding electrostatic potential is

$$\varphi(r) = (e/r) - (e/2R)[(2R/r) + 1]\exp(-r/R). \qquad (1\text{-}9)$$

In the vicinity of $r = 0$ the electrostatic potential is as follows:

$$\varphi(r) \approx (e/2R) - (e/12R^3)r^2 + \dots \,. \qquad (1\text{-}10)$$

Kinetic energy of the electron is a quantum-mechanically averaged value of its operator:

$$T = -(\hbar^2/2m)\Delta. \qquad (1\text{-}11)$$

It follows from Eqs. (1-6), and (1-11) that

$$\langle T \rangle = \hbar^2/8mR^2. \qquad (1\text{-}12)$$

In Quantum Mechanics there is no self-action. Hence, the electrostatic energy of an elementary particle does not exist in Quantum Mechanics. In the simple problem of a hydrogen atom in a ground state there is no electrostatic energy of the electron and the proton. The only electrostatic energy that exists really is the energy of electron-proton interaction. This energy is negative, because of the opposite charges of the interacting particles. So hydrogen atom is a very interesting object, which has electrostatic field and negative electrostatic energy. Such a situation is never possible in Classical Electrostatics. In Classical Electrostatics the energy stored in a total electric field is proportional to the integral of this total electric field in square over the whole space, and it is always positive.

The total energy of a hydrogen atom (without self-action), which consists of the kinetic energy of the electron, Eq. (1-12), and the electrostatic interaction energy of the positively charged core with the potential in the centre of the electronic shell, Eq. (1-10), is as follows:

$$E_h(R) = (\hbar^2/8mR^2) - (e^2/2R). \qquad (1\text{-}13)$$

As was mentioned above, the electrostatic energy of the electron itself should not be taken into account as it represents the so-called self-action. The same should be said about the electrostatic energy of the proton. Energy, described by Eq. (1-13) has a minimum at $R = R_e = \hbar^2/2me^2$, as is well known [1]. Correct minimum value of the ground state energy is given by Eq. (1-7). Corresponding value of the electrostatic energy is $U_0 = -(e^2/2R_e) = -(me^4/\hbar^2)$. This value is the correct one. Indeed, as $E_{0h} = U_0 + \langle T \rangle_0$, $U_0 = E_{0h} - \langle T \rangle_0$, $\langle T \rangle_0 = (me^4/2\hbar^2)$ (as was mentioned above), and $E_{0h} = -(me^4/2\hbar^2)$ (see Eq. (1-7)), we have $U_0 = -(me^4/\hbar^2)$.

Here we have calculated the energy of the ground state of a hydrogen atom in a special way, regarding all the terms of the electrostatic and kinetic energy. We did it on purpose. Thus we see explicitly, that the energy of the ground state of a hydrogen atom, calculated here without taking into account the electrostatic energies of the electron and the proton (that is, without the self-action), has a correct well known and measured experimentally value [1].

A problem of negative hydrogen ion was considered in [3].

Positronium

Now let us imagine a positronium (a hydrogen type atom consisting of electron and positron) in a ground state. The masses of both particles are of the same value. Their electric charges are of the same value, but of the opposite signs. Because of this the wave functions of both particles in a ground state are identical. So here we have no local charge, no electrostatic field but we have negative electrostatic energy of a system.

Uniformly Charged Spherical Ball Model

Let us consider in this section N electrons with the electric charge of each one distributed uniformly in a spherical ball of a radius R. Electrostatic energy of such a ball in Classical Electrostatics is [4] $U = 0.6e^2N^2/R$. Electrostatic energy of each of the electron is $U = 0.6e^2/R$. Electrostatic energy of N electrons is $U = 0.6e^2N/R$. In Quantum Mechanics this energy should be subtracted from the classical energy of a system. So, the electrostatic energy of such a ball in Quantum Mechanics is [3]:

$$U = 0.6e^2N(N-1)/R. \qquad (1\text{-}14)$$

Self-action exists in Classical Electrostatics. So we have N^2 instead of $N(N-1)$ in Equation (1-14) in Classical Electrostatics.

For a spherical ball of a radius R, containing N protons with uniformly distributed over the ball electric charge of each one we have the same equation, Eq. (1-14), for the electrostatic energy.

Let us consider now a spherical ball of a radius R, containing N electrons and N protons, each one of them distributed uniformly inside the ball. The total charge in any point of a ball is zero. So there is no electrostatic field. The total electrostatic energy of such a ball consists of the electrostatic energy of electrons, Eq. (1-14), electrostatic energy of protons, Eq. (1-14), and interaction energy, $U_i = -1.2e^2N^2/R$. The last one is the only one, which should be taken into account. So the total electrostatic energy is described by the following expression:

$$U_t = -1.2e^2N/R. \qquad (1\text{-}15)$$

This result is a particular case of more general expressions, published in [4].

In Classical Electrostatics, as was mentioned, we have N^2 instead of $N(N-1)$ in Equation (1-14). Hence we have the total electrostatic energy zero. In Quantum Mechanics, where self-action does not exist, we have negative total electrostatic energy in a locally electrically neutral system. This occurs because every particle produces electrostatic field and in spite of the local neutrality we should subtract from zero the total electrostatic energy of each separate particle. Thus we come to the negative value of the electrostatic energy.

In this section we have considered a case of no local electric charge, no local electrostatic field and a considerable negative electrostatic energy.

Discussion

The concept of the kinetic energy in non-relativistic Quantum Mechanics is considered here. Kinetic energy is a non-zero positive value for many cases with no visible motion and flux. The momentum operator contains a factor of imaginary unity i. So when a wave function is a real-valued one, the quantum-mechanically averaged momentum, $\mathbf{p}$, is imaginary, $\mathbf{p} = -i\hbar \int \psi \nabla \psi dv$. But using the expansion of the real-valued wave function into Fourier integral, Eq. (1-3), we have $\mathbf{p} = \hbar \int \mathbf{k} a*_{\mathbf{k}} a_{\mathbf{k}} d\mathbf{k}$, which is a real-valued quantity. From this follows that for the real-valued wave function $\mathbf{p} = 0$, as it is the only possibility to be real and imaginary at the same time.

The existence of a positive kinetic energy in the states with no flux can be understood, using expansion in virtual plane waves. This expansion presents the kinetic energy as a sum of partial kinetic energies of constituting virtual plane waves. The real-valued wave function is a superposition of moving plane-waves. Their velocities cancel, but their kinetic energies add.

The energy of the ground state of a hydrogen atom is calculated in a special way, considering all the terms of the electrostatic and kinetic energy of the electron and the proton. It is shown that the correct values of the energy of the ground state and the radius of decay could be obtained only when the electrostatic energy of the electron and the proton are not taken into account. This supports once more the concept that self-action should not be taken into account in Quantum Mechanics.

When self-action in quantum mechanics is not taken into account we have some cases of zero local charge, but non-zero negative electrostatic energy. This occurs, in particular, in a positronium and in the model of a spherical ball, consisting of an equal number of positive and negative elementary particles. This model is also considered in this section.

It should be mentioned here also that as kinetic energy is a function of momentum, and potential energy is a function of coordinates, the uncertainty principle [5] is applicable to these two complementary quantities. This means that the precision with which kinetic and electrostatic energies of an elementary particle could be known simultaneously is fundamentally limited. As follows from Schrödinger equation (see Eq. (1-1)), quantum mechanically averaged sum of kinetic and electrostatic energies is equal to the total energy E, which is the only value, which can be known precisely in Quantum Mechanics.

Conclusions

Simple methodical approach of expanding wave function into a series of virtual plane waves occurred to be helpful for understanding the nature of the kinetic energy of a particle, described by real-valued wave function.

It was shown also that the energy of the electric field does not have to be of a positive value in any imaginable problem.

References

1. LD Landau, and EM Lifshitz, *Quantum Mechanics*, Pergamon Press, Oxford, 1987.

2. Y Kornyushin, *Facta Universitatis: Series PCT*, **2**, N 5, p. 253, 2003.

3. MY Amusia, and Y Kornyushin, *European Journal of Physics*, **21**, N 5, p. 369, 2000.

4. MY Amusia, and Y Kornyushin, *Contemporary Physics*, **41**, N 4, p. 219, 2000.

5. W Heisenberg, *Zeischrift für Physik*, **43**, N 3-4, p. 172, 1927.

2. General Introduction to Low-Dimensional Quantum Movements in Many-Body System

Introduction

Fermi and kinetic energies are basic concepts in many-body systems of free identical particles at low temperature, when free identical particles are degenerate. In Solid State Physics a 3-dimensional case is usually studied [1]. Now 2- and 1- dimensional systems, like thin films and nanowires made of semiconductors, graphenes, carbon nanotubes, nanofibers and others are being studied extensively.

These systems are rather promising in many branches of research and technology. Quantum movements in 2- and 1- dimensional systems of free identical particles are different from those in a 3-dimensional one. Fermi and kinetic energies in 2- and 1-dimensional systems of free identical particles are not the same as in a 3-dimensional one. The purpose of this section is to derive expressions for Fermi and kinetic energies in 2- and 1- dimensional systems of free identical particles, and to obtain criteria of 2- and 1- dimensional quantum movements of free identical particles in a many-body system. These criteria restrict not only dimensions of a sample, but also restrict temperature range.

To perform calculations of Fermi and kinetic energies in a degenerate system of free identical particles the zero boundary conditions model (ZBCM), which was formulated in [2], is used in this section. One should be aware that the calculations performed in the present section are not the exact ones, but they are done in the frame of proposed ZBCM. So obtained results are reliable as far as the ZBCM is relevant.

Fermi and Kinetic Energies

To calculate Fermi and Kinetic energies the periodic boundary conditions model (PBCM) is used as a rule [1]. This model is rather adequate for a bulk solid. But when the motion of the particles is confined, the PBCM does not look like a proper one. More adequate in this case is the potential box model. When the potential box is infinitely deep, zero boundary conditions on the borders of the box are realized, thus leading to ZBCM.

Fermi and Kinetic Energies in a 3-Dimemsional ZBCM

In ZBCM an ensemble of free identical particles, confined in an infinitely deep potential box, is considered [2]. Schrödinger equation for the wave function of a particle $\psi(x,y,z)$ is (provided the value of the potential in the box is zero):

$$-(\hbar^2/2m)\Delta\psi(x,y,z) = E\psi(x,y,z), \qquad (2\text{-}1)$$

where m is the mass of a particle and $\hbar$ is the Planck constant divided by 2π.

The wave functions are described by the following equation [2]:

$$\psi(x,y,z) = (8/abc)[\sin(\pi v_x x/a)][\sin(\pi v_y y/b)][\sin(\pi v_z z/c)]. \ v_i = 1, 2, 3, \dots . \ (2\text{-}2)$$

The components of the wave vectors are [2] $k_i = \pi v_i/a_i$ (here a, b and c are the dimensions of a box and $i = x, y, z$; $a_x = a$, $a_y = b$, and $a_z = c$).

The energy levels are described by the following relation [2]:

$$E(v) = (\pi^2\hbar^2/2m)[(v_x/a)^2 + (v_y/b)^2 + (v_z/c)^2].$$ (2-3)

The energy of a ground state ($v_i = 1$) is not zero: $E_0 = (\pi^2\hbar^2/2m)[(1/a^2) + (1/b^2) + (1/c^2)]$, but it is rather small when a, b, and c are quite large. The energy levels in ZBCM are:

$$E(v) = (\pi^2\hbar^2/2mV^{2/3})v^2,$$ (2-4)

where $V = abc$ is the volume of a sample, and $v^2 = [(V^{1/3}/a)v_x]^2 + [(V^{1/3}/b)v_y]^2 + [(V^{1/3}/c)v_z]^2$.

When $N = nV$ confined particles fill in the levels (for spin 1/2 each level is occupied by two particles), they reach in the process of the filling the limiting value of v, v_F. As v_i are positive numbers only (see Eq. (2-2)), one has only 1/8 fraction of the incomplete sphere (in the v_i space, excluding planes corresponding to $v_i = 0$) of a radius v_F to be filled in. This yields the total number of the states filled being equal to $(\pi/3)v_F^3 - 0.75\pi v_F^2$ (as $v_i > 0$ only, and $v_i = 0$ does not exist as was pointed out above). They are all occupied by the particles in consideration. For N much larger than unity (which is a regular case) one has

$$(\pi/3)v_F^3 = nV, \text{ or } v_F = (3nV/\pi)^{1/3}.$$ (2-5)

Equations (2-4), and (2-5) yield for the Fermi energy the following expression:

$$E_F = E(v_F) = (\hbar^2/2m)(3\pi^2 n)^{2/3}.$$ (2-6)

Equation (2-6) yields a well-known result, obtained in PBCM [1].

When one has η particles in a unit volume, the Fermi energy of a system is $E_F = (\hbar^2/2m)(3\pi^2\eta)^{2/3}$. If $d\eta$ particles are added, the increase in the energy of a system is $(\hbar^2/2m)(3\pi^2\eta)^{2/3}d\eta$. As the potential energy in a box is assumed to be zero, the kinetic energy of a system is equal to the total energy. The total energy (per unit volume) is the integral of the last expression on η from 0 to n. So, for the kinetic energy, T_k, we have:

$$T_k = 0.6NE_F. \qquad (2\text{-}7)$$

Equation (2-7) is also a well-known result obtained in PBCM [1].

Fermi and Kinetic Energies in 2-Dimensional ZBCM

In 2-dimensional case Eq. (2-4) should be written as follows [2]:

$$E(v) = (\pi^2\hbar^2/2mS)v^2, \qquad (2\text{-}8)$$

where $S = ab$ and $v^2 = [(S^{1/2}/a)v_x]^2 + [(S^{1/2}/b)v_y]^2$.

In 2-dimensional case only 1/4 part of the disk (in the v_i space) is relevant because v_i are positive numbers only. Then, instead of $(\pi/3)v_F^3 = nV$, one has $(\pi/2)v_F^2 = N = n_sS = nV$ (n_s is the number of the confined particles per unit area). Instead of $v_F = (3nV/\pi)^{1/3}$ one has $v_F = (2n_sS/\pi)^{1/2}$. One should also take into account that the third (neglected) dimension of a size $\delta \to 0$ yields (according to Eq. (2-3)) the following contribution to the energy of a system: $E(\delta) = (\pi\hbar)^2/2m\delta^2$. So the Fermi energy is

$$E_F = E(v_F) + E(\delta) = (\hbar^2/2m)[2\pi n_s + (\pi/\delta)^2]. \qquad (2\text{-}9)$$

Equation (2-9) was also derived for the case of N much larger than unity. The term π/δ^2 in Eq. (2-9) is a very essential one when δ is small.

Calculation of the kinetic energy in 2-d case like it was done previously yields

$$T_k = (\hbar^2/2m)[\pi N n_s + (\pi/\delta)^2 N].\qquad(2\text{-}10)$$

The last term (proportional to N/δ^2) in the right-hand part of Eq. (2-10), is a very essential one when δ is small.

Fermi and Kinetic Energies in 1-Dimensional ZBCM

In 1-dimensional case Eq. (2-4) should be written [2] as

$$E(v) = (\pi^2\hbar^2/2ma^2)v^2.\qquad(2\text{-}11)$$

In 1-dimensional case only half of the double length (in the v space) is relevant because of the positive v numbers only. Then, instead of $(\pi/3)v_F^3 = nV$ one has $2v_F = N = n_l l$ (n_l is the number of the confined particles per unit length), so $v_F = 0.5n_l l$, and

$$E_F = E(v_F) + E(\delta) + E(\delta_{max}) = (\hbar^2/2m)[(\pi n_l/2)^2 + (\pi/\delta)^2 + (\pi/\delta_{max})^2].\quad(2\text{-}12)$$

Equation (2-12) was also derived here for the case of N much larger than unity. Two last terms in the right-hand part of Eq. (2-12) represent the contribution of the neglected dimensions to the energy of a system (δ_{max} is the larger size of the neglected dimensions).

The kinetic energy in the 1-dimensional case, calculated as previously, is

$$T_k = (\pi^2\hbar^2/2m)N[(n_l^2/12) + (1/\delta^2) + (1/\delta_{max})^2].\qquad(2\text{-}13)$$

The contribution of the neglected dimensions is a very essential one when δ and/or δ_{max} are/is small.

In a Potential Box of a Finite Depth all the Energy Levels (Including Fermi Energy) are Lower than those in the Box of an Infinite Depth

The equations for wave vectors k_i in a finite depth potential box are as follows [3]

$$k_i a_i + 2\arcsin[\hbar k_i/(2mU)^{1/2}] = v_i\pi, \ v_i = 1, 2, 3, \dots, \qquad (2\text{-}14)$$

where U is the potential depth.

As Eq. (2-14) has an additional (comparative to the case of the model of infinitely deep potential box) positive term in the left-hand part of it, one can see that the corresponding k_i and $E(v_i) = (\hbar k_i)^2/2m$ are smaller than these in the model of infinitely deep potential box either in PBCM, or in ZBCM. The model of the potential box of a finite depth yields lower (comparative to the case of the model of the infinitely deep potential box) energy levels, and the Fermi energy in this model is also lower.

It is worthwhile to note also that in the model of the finite depth of a potential box the boundary conditions are non-zero. From this follows that the wavelength of each state in this model is larger (comparative to that in the infinitely deep potential box model), and corresponding wave vectors and energies are smaller.

Criteria of Low-Dimensional Quantum Movements

Criterion of 2-Dimensional Quantum Movement

Let one of the dimensions be small: $c = \delta \to 0$. Then the first excited energy level ($v_z = 2$), corresponding to this direction in the ZBCM is $2\pi^2\hbar^2/m\delta^2$ (see Eq. (2-3)). When this quantity is larger than the 2-d Fermi energy (see Eq. (2-9)),

$$2\pi^2\hbar^2/m\delta^2 > (\hbar^2/2m)[(2\pi n_s) + (\pi/\delta)^2], \qquad (2\text{-}15)$$

the quantum movement could be considered as a 2-d one.

Using Eq. (2-15) and taking into account that the surface density of the particles n_s = $n\delta$, one can write

$$\delta < (3\pi/2n_s)^{1/2}, \text{ or } \delta < (3\pi/2n)^{1/3}. \qquad (2\text{-}16)$$

From Eq. (2-16) follows that the density of the delocalized particles n is the most important parameter in the problem considered. Thus for semiconductor with n = $(3\pi/2)\times10^{15}$ cm^{-3} we have δ < 100 nm. One should say that it is quite a considerable size, much larger than typical sizes of regular nanoobjects. So quantum movements of delocalized electrons in nano-thin films made of semiconductors are often of a 2-dimensional character. Carbon nanoobjects like fullerene molecules, carbon nanotubes, carbon peapods, and graphenes usually have 4 delocalized electrons per one carbon atom [4]. That is the density of the delocalized electrons in carbon nanoobjects n is about 10^{24} cm^{-3}, which is two orders of magnitude larger than that in metals, where n is about 10^{22} cm^{-3}. According to Eq. (2-16) the size of a carbon nano-sample should be less than 0.1 nm to perform low-dimensional properties of quantum movement of delocalized electrons. So for regular nanoobjects with sizes larger than 0.1 nm the quantum movements of the delocalized electrons are always essentially 3-dimensional ones.

Criterion of 1-Dimensional Quantum Movement

When two of the three dimensions are small, and the first excited energy level (corresponding to v = 2) of the largest of the small dimensions δ_{max} is $2\pi^2\hbar^2/m(\delta_{max})^2$, and it is larger than 1-d Fermi energy (see Eqs. (2-3), and (2-12)),

$$2\pi^2\hbar^2/m(\delta_{max})^2 > (\hbar^2/2m)[(\pi n_l/2)^2 + (\pi/\delta)^2 + (\pi/\delta_{max})^2], \qquad (2\text{-}17)$$

quantum movement can be considered as a 1-d one.

From Eq. (2-17) follows that

$$\delta_{max} < [12/(4 + \delta^2 n_l^2)]^{1/2}\delta. \qquad (2\text{-}18)$$

Taking into account that the linear density of the particles $n_l = n\delta\delta_{max}$, one can see that for the case $\delta_{max} = \delta$, Eq.(2-18) yields:

$$\delta < 2^{1/2}/n^{1/3}. \qquad (2\text{-}19)$$

For $n = 2^{3/2}\times10^{15}$ cm^{-3} we have $\delta < 100$ nm. This is a very considerable value, much larger than one manometer. Here also the one-dimensional quantum movement of the delocalized electrons could be realized in nanofibers made of semiconductors and not of metals, like in before. Quantum movements of delocalized electrons in carbon nanotubes are always essentially 3-dimensional ones.

Temperature Criterion of Low-Dimensional Quantum Movements

Low-dimensional movements are realized when the difference between the first excited energy level, corresponding to the small direction in ZBCM, $2\pi^2\hbar^2/m\delta^2$ (see Eq. (2-3)) for the corresponding value of $v = 2$), and the ground state level, $\pi^2\hbar^2/2m\delta^2$ (see Eq. (2-3) for the corresponding value of $v = 1$), is essentially larger then kT (k here is the Boltzmann constant and T is temperature). From this follows that temperature T should be much lower than $3\pi^2\hbar^2/2km\delta^2$:

$$T \ll 3\pi^2 \hbar^2/2km\delta^2. \qquad (2\text{-}20)$$

For $\delta = 10^{-6}$ cm and $m = 9.1 \times 10^{-28}$ g Eq. (2-20) yields that T should be much lower than 131 K. This value is not an extreme one.

Discussion

It is shown here that the ZBCM is rather inherent for the description of free charged particles in a potential box. Obtained results are quite close to those obtained in the PBCM [1]. Criteria of realization of low-dimensional movements are derived, including restriction on the temperature. At a comparatively high temperature the quantum movements of particles can be never considered as low-dimensional ones. At low enough temperatures every concrete object that looks like low-dimensional one should be checked according to simple criteria described in this sedction if the movements of the particles in it is really low-dimensional ones.

Quantum movements of delocalized electrons in regular carbon nanoobjects are usually of a 3-dimensional character [4].

It is worthwhile to mention here that some collective movements in carbon nanoobjects are of a classical origin, and they don't have to be low-dimensional ones [4].

References

1. C Kittel, *Introduction to Solid State Physics*, New York, Wiley, 1996.
2. Y Kornyushin, *Facta Universitatis: Series PCT*, **2**, N 3, p. 141, 2001.
3. LD Landau, and EM Lifshitz, *Quantum Mechanics*, Oxford, Pergamon, 1984.
4. Y Kornyushin, *LTP*, **34**, N 10, p. 838, 2008.

3. Quantum Rotator: Some Examples in Modern Spirit

Introduction

Rotating body is a subject of Quantum Mechanics [1]. We regard here rotations around principal axes of a body [1 - 3]. In general case there are 3 principal moments of inertia of a body [1 - 3]. Quantum levels of rotator are [1]

$$E_J = (\hbar^2/2I)J(J+1). \tag{3-1}$$

Here $\hbar$ is Planck constant divided by 2π, principal moment of inertia of a body [2] $I = \int\rho(\mathbf{r})x^2 dV$ (integral is taken over the body volume V, $\rho(\mathbf{r})$ is the mass density, and x is the perpendicular distance to the principal axis of rotation); J is zero or integer.

When the mass density is constant along the body, one can see that the moment of inertia of a body is $I = m\langle x^2 \rangle$ (here m is the mass of a body, $\langle \ldots \rangle$ denotes averaging over the volume of a body).

Now let us consider a body of a shape of a spherical layer, $R_i \leq r \leq R$. For this spherically symmetric object there is only one principal moment of inertia. Calculating it, $I = \int\rho(\mathbf{r})x^2 dV$, one gets

$$I = 0.4mR^2 + 0.4mR_i^3[(R + R_i)/(R^2 + RR_i + R_i^2)]. \tag{3-2}$$

We shall consider further on several rotating objects.

Atoms and Ions

Atom (ion) consists of a nucleus and an electronic cloud. We shall consider a nucleus in the nuclear liquid drop model [4]. We assume that the density of a mass in the nucleus is constant in the sphere of a radius R. It follows from Eq. (3-2) that, as $R_i = 0$ for the nucleus, $I = 0.4mR^2$ and

$$E_J = 1.25(\hbar^2/mR^2)J(J+1) = 1.25(\hbar^2/mr_0^2A^{2/3})J(J+1). \qquad (3\text{-}3)$$

For the nucleus of a carbon atom $m = 2\times10^{-26}$ kg [5]. The radius of a nucleus $R = r_0A^{1/3}$ [1] (here A is the number of the nucleons in the atom, $r_0 = 1.2\times10^{-15}$ m). For carbon $A = 12$ and according to Eq. (3-3) $E_J = 0.5754J(J+1)$ MeV. This is an enormous value, inaccessible. No rotation is practically possible.

Let us consider electron cloud of the atom/ion in the nuclear liquid drop model also [6]. In this model the radius of the electronic cloud of the atom/ion is as follows [6]

$$R_e = 7.37[N^{2/3}/(5Z - 2N)](g\hbar^2/m_ee^2). \qquad (3\text{-}4)$$

Here N is the number of the electrons in the cloud, Z is the number of the protons in the nucleus, $g = 0.317$ [6], and m_e is the electron mass. For the atom/ion Eqs. (3-1), (3-2), and (3-4) yield:

$$E_J = 1.25(\hbar^2/Nm_eR_e^2)J(J+1) = 0.023[(5Z - 2N)^2/g^2N^{7/3}](m_ee^4/\hbar^2)J(J+1). \quad (3\text{-}5)$$

For a neutral atom of carbon $N = Z = 6$ and we have $E_J = 6.8056(m_ee^4/\hbar^2)J(J+1)$. It is worthwhile to note that $m_ee^4/\hbar^2 = 27.21$ eV [1]. The value of E_J for the electronic

25

cloud is also of a very large number. So rotation of this object is hardly possible.

Fullerene Molecule

For the delocalized electron cloud of a fullerene molecule (C_{60}) $N = 240$, $R_i = 0.279$ nm, and $R = 0.429$ nm [7]. For these numbers Eq. (3-2) yields $I = 21.5364m_e$ (in $m_e$nm^2 units). For these numbers Eq. (3-1) yields $E_J = 0.001771J(J + 1)$ eV. For $J = 1$ we have $E_J = 0.003542$ eV. This value corresponds to 41.1 K. When temperature is essentially smaller than 40 K the behavior of this rotator has a quantum character. Anyway rotation of the electronic cloud of a fullerene molecule is fairly possible.

The ionic core of a fullerene molecule has radius $R_f = 0.354$ nm [7]. Taking into account that as was mentioned before the mass of a carbon atom is 2×10^{-26} kg, and there are 60 carbon atoms in a fullerene molecule, we have $E_J = 2.31 \times 10^{-7}J(J + 1)$ eV. For $J = 1$ we have $E_J = 4.62 \times 10^{-7}$ eV. This corresponds to 0.005362 K. This quantity is so small that quantum behavior is utterly impossible. This object rotates always classically.

Ring of a Nanotube

Let us consider a ring of a nanotube of a radius R. This ring has a length $2\pi R$. We assume that the radius of a ring R is essentially larger than the diameter of a nanotube d (d is typically about 1.4 nm [8]). We accept also as in [8] that the number of the delocalized electrons in a nanotube is $N_n = 670$ nm^{-1}. Then Eq. (3-1) yields

$$E_J = (\hbar^2/4\pi m_e N_n R^3)J(J + 1). \qquad (3\text{-}6)$$

For $R = 100$ nm we have $E_J = 2.846 \times 10^{-11}$ eV. This is an extremely small quantity. This means that rotation of this object is fairly classical. The same should be said about rotation of the ion core of the regarded ring, and rotation of the ring as a whole around the axis, going through the middle of the ring.

References

1. LD Landau, and EM Lifshitz, *Quantum Mechanics*, Pergamon, Oxford, 1987.

2. H Goldstein, *Classical Mechanics*, Addison-Wesley, Cambridge Mass., 1950.

3. JJ Sylvester, *Phil. Mag.*, **4**, N 23, p. 138, 1852.

4. A Bohr, and BR Mottelson, *Nuclear Structure, V 2*, Benjamin, London, 1975.

5. RC Weast (ed.), *1988 CRC Handbook of Chemistry and Physics*, CRC, Boca Raton, FL.

6. MY Amusia, and Y Kornyushin, *Contemporary Physics*, **41**, N 4, p. 219, 2000

7. A Rüdel, R Hentges, U Becker, HS Chakraborty, ME Madjet, and JM Rost, *Phys. Rev. Lett.*, **89**, N 12, p. 125503, 2002.

8. Y Kornyushin, *LTP*, **34**, N 10, p. 838, 2008.

4. Simplified Theory of Polarons and Bipolarons

Introduction

Localized states of current carriers in crystals (polarons) have been known since the 1940s [1,2]. The concept of localized states plays an important role in many areas of condensed matter physics, including superconductivity [3,4]. The concept of a polaron was introduced by SI Pekar [2] and subsequently developed by many authors.

A polaron is a quasiparticle that interacts with the polarization oscillations of a crystal lattice (especially ionic crystals) such that an autolocalized state of the current carriers arises. An autolocalized state is a particular case of a localized state and is qualitatively different from the free state of a particle. Autolocalization occurs in a homogeneous medium due to the internal properties of the medium. The phenomenon is different from localization in an external potential.

The existing theory of auto localization [1-4] is a rather developed and successful theory. But its study requires advanced mathematics, including the calculus of variations. Moreover, the theoretical calculations are very tedious, complicated, and must be done numerically. Instead, we will present a simple model for the autolocalization of a free charged particle. This model can be used to introduce the study of auto localization and help students understand the basic concepts of the subject without performing complicated calculations. The polarization potential well in the proposed model is deep enough for only one localized level. In a more realistic theory several localized levels can be found in a self-consistent potential well.

In dielectric materials with a sufficiently large dielectric constant, two identical charged particles can be localized in one polarization potential well, forming a bipolaron. A bipolaron is also a well-known phenomenon [2-4] and is thought to be a carrier of a superconducting current in many materials [3]. The polarization potential well for a bipolaron in the simple model that we will discuss is deep enough for only one localized level, in contrast to more accurate theories in which several levels can exist [2-4].

When a charged particle is in a localized state, an electrostatic field arises around it. This field polarizes the surrounding medium, so that every small volume of the medium acquires a small dipole moment. These moments are oriented so that there is an attraction between the moments and a particle in a localized state, which means that the interaction energy is negative. So a charged particle in a localized state creates a potential well for itself in which it could be localized if the well is deep enough. If the well is not deep enough, the particle will not be in a localized state.

Let us consider a particle with a wave function $\psi(\mathbf{r})$ and a charge e. The charge density is $\rho(\mathbf{r}) = e\psi^*(\mathbf{r})\psi(\mathbf{r})$ [5]. Let the corresponding electrostatic potential in the dielectric medium with a dielectric constant ε be $\varphi_0(\mathbf{r})/\varepsilon$, where $\varphi_0(\mathbf{r})$ is the electrostatic potential of the same particle in a vacuum [6]. The potential of the charged particle in a dielectric medium can be written as

$$\varphi_0(\mathbf{r})/\varepsilon \equiv \varphi_0(\mathbf{r}) - [(\varepsilon - 1)/\varepsilon]\varphi_0(\mathbf{r}). \qquad (4\text{-}1)$$

The first term on the right-hand side of Eq. (4-1) represents the potential of the particle only, and the second term represents the potential produced by the polarization of the dielectric medium. Thus, the second term of the right-hand side of Eq. (4-1), $\varphi_d(\mathbf{r}) = -[(\varepsilon - 1)/\varepsilon]\varphi_0(\mathbf{r})$, is the part of the potential that acts on the particle. This potential is an attractive one, because ε is always larger than unity.

Autolocalized Charged Particle

To determine the possibility of autolocalization, we assume the following model for the localized wave function and hence for the charge density:

$$\psi(r) = (g^{3/2}/\pi^{3/4})\exp[-0.5(gr)^2], \qquad (4\text{-}2a)$$

$$\rho = e(g^3/\pi^{3/2})\exp[-(gr)^2]. \qquad (4\text{-}2b)$$

The quantity $1/g$ in Eqs. (4-2) represents the autolocalization radius, because for r larger than $1/g$, the density of the particle vanishes. The proposed model represented by Eqs. (4-2) was chosen because the wave function in Eq. (4-2a) is similar to those arising in a smooth potential well. Eq. (4-2b) is an exact consequence of Eq. (4-2a) [5]. The total electric field, calculated by using Gauss's theorem, is

$$E(r) = (e/\varepsilon r^2)\Phi(gr) - [2eg/(\pi^{1/2}\varepsilon r)]\exp{-(gr)^2}. \qquad (4\text{-}3)$$

Here $\Phi(x)$ is the probability integral, that is, the integral of $[(2/\pi^{1/2})\exp{-y^2}]$ from $y = 0$ to $y = x$. The part of the field that acts on the particle produced by the by the polarization of the dielectric medium is

$$E_d(r) = [(\varepsilon - 1)/\varepsilon]\{[2eg/(\pi^{1/2}r)][\exp{-(gr)^2}] - (e/r^2)\Phi(gr)\}. \qquad (4\text{-}4)$$

The first term of the expansion in r of the right-hand part of Eq. (4-3) yields

$$E(r) = (4eg^3/3\pi^{1/2}\varepsilon)r + \dots \ . \qquad (4\text{-}5)$$

As follows from Eq. (4-5), the electrostatic potential $\varphi(\mathbf{r})$ is

$$\varphi(r) = \varphi(0) - (2/3\pi^{1/2})(eg^3/\varepsilon)r^2 + \dots \ , \qquad (4\text{-}6)$$

with the part produced by the polarization that acts on the particle given by

$$\varphi(r) = [(\varepsilon - 1)/\varepsilon][(2/3\pi^{1/2})eg^3r^2 - \varepsilon\varphi(0)] + \dots \ . \tag{4-7}$$

This approximate potential corresponds to a three-dimensional harmonic oscillator. The model wave function assumed in Eq. (4-2a) corresponds to the ground state of the three-dimensional harmonic oscillator. The potential represented by Eq. (4-7) is formed by many electrons and ions. For this reason we can that the total mass of all the particles whose charges form the polarization potential (represented by Eq. (4-7)) is much larger than the mass of the particle under investigation. Hence, it is possible to approximate the reduced mass by the original value of the mass of a particle. This point is a crucial one for the phenomenon of autolocalization. Only the polarization of a medium created by heavy enough particles can lead to the autolocalization of a charged particle, because the structure of the potential, which keeps the particle localized, should be relatively stable with respect to the motion of the localized particle.

An auto localized current carrier in a crystal together with the surrounding polarization of the medium created by heavy inert particles is usually called a polaron [1-4]. When the motion of a polaron is considered, we have to take into account that the motion of the surrounding cloud of polarization charges accompanies the motion of the localized particle. This phenomenon is expressed by the larger effective mass of a polaron compared to the mass of a bare particle. Hence, the polaron current carriers have low mobility and the sample has high electrical resistance. Now let us calculate the depth of the potential well $\varphi(0)$. For $r \to \infty$ the potential $\varphi(r)$ goes to zero, which means that the value of $\varphi(0)$ is equal to the integral of $E(r)$ from zero to infinity. Hence, we have $\varphi(0) = 2eg/\pi^{1/2}\varepsilon$. The electrostatic potential energy of a charged particle is equal to its charge multiplied by the potential due to the polarization medium, and thus from Eq. (4-1) it is given by

$$W(0) = -(2/\pi^{1/2})[(\varepsilon - 1)/\varepsilon]e^2 g \qquad (4\text{-}8)$$

The energy of a particle is, according to Eqs. (4-7) and (4-8),

$$W(r) = (2e^2/\pi^{1/2})[(\varepsilon - 1)/\varepsilon][(g^3/3)r^2 - g] + \ldots$$
$$= -(2/\pi^{1/2})[(\varepsilon - 1)/\varepsilon]e^2 g + 0.5m\omega^2 r^2 + \ldots \qquad (4\text{-}9)$$

where m is the original effective mass of the particle and ω is the angular frequency of a harmonic oscillator, which is given by

$$\omega = (2/3^{1/2}\pi^{1/4})[(\varepsilon - 1)/\varepsilon]^{1/2}(e/m^{1/2})g^{3/2}. \qquad (4\text{-}10)$$

In the ground state the energy of the three-dimensional oscillator is $1.5\hbar\omega$, and it is smaller than the depth of the potential well, $W(0)$. From this inequality follows that

$$g < (4/3\pi^{1/2})[(\varepsilon - 1)/\varepsilon](me^2/\hbar^2) \qquad (4\text{-}11)$$

Equation (4-11) expresses the condition for the autolocalization of the charged particle in a dielectric medium.

The Equilibrium Value of the Inverse Autolocalization Radius

The equilibrium value of the inverse autolocalization radius, $1/g$, is determined by the minimum of the energy of the system. The system consists of a charged particle and the polarized dielectric around it. Let us consider now the intrinsic energy of the polarization itself. The density of the electrostatic energy is the product of $\mathbf{D}$ and $\mathbf{E}$ divided by 8π. The part of the induction, $\mathbf{D}$, due to the polarization $\mathbf{D}_p$, is determined by the equation $\mathrm{div}\mathbf{D}_p = 0$. Because we are considering the intrinsic energy of the polarization, we should not include the charge of the particle. Because $\mathbf{D}_p$ has a radial component only, we have $\mathrm{div}\mathbf{D}_p = (dD_p/dr) + (2/r)D_p = 0$. The only solution of this equation, $D_p = C/r^2$ (C is a constant), goes to infinity at $r = 0$ when $C \neq 0$, which is not acceptable. Thus we conclude that $D_p = 0$. This solution corresponds to the absence of a free charge and a presence of a bound charge [6]. The value of the electric field is not zero, but $D_p = E_p + 4\pi P = 0$ (E_p is the electric field due to the polarization, and P is the density of the electric dipole moment of the dielectric, which is formed by the bound charge). Because $D_p = 0$, the intrinsic energy of the polarization is also zero, even though $E_p = -4\pi P \neq 0$.

The intrinsic self-interaction energy of the particle should be excluded. What is left is the energy of the interaction of the particle with the polarized medium induced by the charged particle. The energy at the bottom of the potential well is $W(0)$ (Eq. (4-8)), but the particle occupies its ground state with energy $1.5\hbar\omega$. Thus, it follows from Eqs. (4-8) – (4-10) that the interaction energy is as follows

$$W_i = (3^{1/2}/\pi^{1/4})[(\varepsilon - 1)/\varepsilon]^{1/2}(e\hbar/m^{1/2})g^{3/2} -$$
$$(2/\pi^{1/2})[(\varepsilon - 1)/\varepsilon]e^2g. \qquad (4\text{-}12)$$

The right-hand side of Eq. (4-12) has a minimum at

$$g = g_e = (16/27\pi^{1/2})[(\varepsilon - 1)/\varepsilon](me^2/\hbar^2). \qquad (4\text{-}13)$$

For typical value $\varepsilon = 5$, $g_e = 0.267(me^2/\hbar^2)$.

At $g = g_e$ the inequality (4-11) yields $(4/9) < 1$. The ratio of $1.5\hbar\omega$ to $W(0)$ at $g = g_e$ is equel to $2/3$, which means that at least one level (the ground state level) could be formed in the potential well. But the energy of the first excited level at $g = g_e$ is $1.5\hbar\omega + \hbar\omega = 2.5\hbar\omega$, which is larger than the depth of the potential well $-W(0)$ (the ratio of $2.5\hbar\omega$ to $-W(0)$ is 10/9). Hence, the potential well in our model is not deep enough to have a first excited level. Moreover, the approximate model potential (which is derived from the model wave function) is not correct for the wave function of the first excited level for a three-dimensional harmonic oscillator.

In a more realistic theory several localized levels can be found often in a self-consistent potential well [2-4]. So the model regarded here is not suitable for the description of such a case.

The Binding Energy

The binding energy is determined from Eqs. (4-12), and (4-13):

$$W_b = W_i(g_e) = -(32/81\pi)[(\varepsilon - 1)/\varepsilon]^2(me^4/\hbar^2). \tag{4-14}$$

For $\varepsilon = 5$ we have $W_b = -0.0805(me^4/\hbar^2) = -2.19$ eV, which corresponds to a temperature of 25415 K, which is much larger than the ambient temperature. Because the depth of the potential well is much larger than kT, all the electrons are in an auto localized state. So in an autolocalized state, the electrons in a conductivity band have considerably lower energy than in a free (plane wave) state. This lower energy influences the concentration of the electrons in a conductivity band, because their concentration depends exponentially on the gap energy [7]. (The gap energy is the difference between the electron energy at the bottom of the conductivity band and at the top of the valence band.) The concentration of the current curriers significantly influences the electrical conductivity and optical properties. Optical properties are in a great extent dependent on the plasma frequency [7], which is proportional to $(n/m_a)^{1/2}$ (n is the number density of the current carriers in a conductivity band and m_a is the effective mass of an electron in an autolocalized state). So autolocalization significantly affects various physical properties.

Current Carriers (Electrons and Holes) in a Semiconductor

The present approach is valid when it is possible to consider a separate localized current carrier and to neglect the presence of the other current carriers. This assumption is valid when the Debye screening radius $1/g_D$ is much larger than the localization radius of the current carrier $1/g_e$. When the current carriers present in the sample are of one type only and are not degenerate [7], which means that the density of the current carriers is not too high and the temperature is not too low, then

$$g_D^2 = 4\pi e^2 n/\varepsilon kT. \qquad (4\text{-}15)$$

We have from Eqs. (4-13) and (4-15) inequality

$$n \ll n_u = 0.00890[(\varepsilon - 1)^2/\varepsilon]kT(m^2 e^2/\hbar^2). \qquad (4\text{-}16)$$

Here n_u is the upper limit of the concentration of the current carriers, defined by Eq. (4-16). On the other hand, to consider the current carriers as noninteracting, we should find much less (on the average) than one current carrier inside the sphere of a Debye screening radius. In this case every current carrier is screened from the influence of the others. Thus we have $4\pi n/3g_D^3 \ll 1$, and it follows that

$$n \gg n_l = (1/36\pi)(\varepsilon kT/e^2)^3. \qquad (4\text{-}17)$$

Here n_l is the lower limit of the concentration of the current carriers, defined by the inequality (4-17). The inequalities (4-16) and (4-17) can be combined to give

$$n_l \ll n \ll n_u \qquad (4\text{-}18)$$

For m equal to the electronic mass, $\varepsilon = 5$, and $T = 300$ K, we have $n_l = 6.41\times10^{15}$

cm^{-3}, and $n_u = 1.82 \times 10^{20}$ cm^{-3}. Because the concentration of the current carriers in semiconductors is about $10^{17} - 10^{18}$ cm^{-3} [6], the inequalities (4-18) are satisfied by the properties of many materials.

Bipolaron

Let us consider two identical particles in a potential well. Two current carriers in a self-consistent potential well are called bipolaron [2-4]. Bipolarons play important role in Solid State Physics, and they can be superconducting current carriers in some materials [3]. Both of the charged particles of a bipolaron occupy the ground state (they could be bosons or fermions with opposite spines in general). Let ψ, ρ, $E(r)$, $E_0(r)$, and $E_d(r)$, be the same as before for each particle. It follows from Eq. (4-9) that the potential for both particles, each one with charge e (the total charge is $2e$), is

$$W(r) = -(4/\pi^{1/2})[(\varepsilon - 1)/\varepsilon]e^2g + (4/3\pi^{1/2})[(\varepsilon - 1)/\varepsilon]e^2g^3r^2 + \dots \ . \qquad (4\text{-}19)$$

The last term on the right-hand side of Eq. (4-19) is equal to $0.5m\omega_b^2r^2$. From this follows that

$$\omega_b = (2^{3/2}/3^{1/2}\pi^{1/4})[(\varepsilon - 1)/\varepsilon]^{1/2}(e/m^{1/2})g^{3/2}. \qquad (4\text{-}20)$$

The direct interaction of the identical particles should be taken into account. It is determined by the integral of $2E_0^2(r)$ over all space divided by 8π. This integral yields

$$W_i = (2/\pi)^{1/2}e^2g. \qquad (4\text{-}21)$$

The energy of the two particles in the ground state of a potential well is $3\hbar\omega_b$. In an autolocalized state the sum of this energy and W_i is smaller than the depth of the potential well $(4/\pi^{1/2})[(\varepsilon - 1)/\varepsilon]e^2g$ (see Eq. (4-19)). From this follows that

$$0 < g^{1/2} < (1/2^{3/2}3^{1/2}\pi^{1/4})[(\varepsilon-1)/\varepsilon]^{1/2}\{[(4-2^{1/2})\varepsilon-4]/(\varepsilon-1)\}(em^{1/2}/\hbar). \quad (4\text{-}22)$$

Equation (4-22) is the condition for the existence of at least one localized level in a self-consistent potential well. Note that $g^{1/2}$ is positive when $\varepsilon > 4/(4-2^{1/2}) \approx 1.55$ only. When $g^{1/2}$ is negative, there is no localization because the wave function is not localized. The direct interaction of the two particles results in an increase of the energy of the system. The attractive interaction of the particles with the polarization of the medium leads to a decrease of the energy. But if ε is not large enough, this decrease is not sufficiently large to compensate for the energy increase due to the repulsion of two particles. In this case the resulting potential well is not deep enough and the ground state cannot be formed. In such materials a bipolaron does not exist.

The total energy of the system of two particles as follows from Eqs. (4-19) - (4-21) is as follows:

$$W(g) = (2^{3/2}3^{1/2}/\pi^{1/4})[(\varepsilon-1)/\varepsilon]^{1/2}(e\hbar/m^{1/2})g^{3/2} +$$
$$(2/\pi)^{1/2}e^2g - (4/\pi^{1/2})[(\varepsilon-1)/\varepsilon]e^2g. \quad (4\text{-}23)$$

The minimum of this energy corresponds to

$$g_e^{1/2} = (1/2^{1/2}3^{3/2}\pi^{1/4})[(\varepsilon-1)/\varepsilon]^{1/2}\{[(4-2^{1/2})\varepsilon-4]/\varepsilon\}(em^{1/2}/\hbar). \quad (4\text{-}24)$$

We can see that $g_e^{1/2}$ is positive for $\varepsilon > 4/(4-2^{1/2}) \approx 1.55$, in accord with the previous result. Equation (4-24) can be rewritten as

$$g_e = (1/54\pi^{1/2})[(\varepsilon-1)/\varepsilon]\{[(4-2^{1/2})\varepsilon-4]/\varepsilon\}^2(me^2/\hbar^2). \quad (4\text{-}25)$$

For $\varepsilon = 5$, Eq. (25) yields $g_e = 0.0416(me^2/\hbar^2)$. This value of the inverse autolocalization radius is much smaller than that calculated from Eq. (4-13), which means that the localization radius of a bipolaron is much larger than that of an autolocalized single charged particle.

The equilibrium energy of the system according to Eq. (4-23) is

$$W(g_e) = -(1/162\pi)\{[(4 - 2^{1/2})\varepsilon - 4]^3/\varepsilon^2(\varepsilon - 1)\}(me^4/\hbar^2). \qquad (4\text{-}26)$$

Again the minimum energy $W(g_e)$ is positive for $\varepsilon > 4/(4 - 2^{1/2}) \approx 1.55$, in accord with our previous result. For $\varepsilon = 5$, the minimum energy $W(g_e) = -0.0140(me^4/\hbar^2)$, which is much smaller than the binding energy of two separate particles. The autolocalized state of two separate particles (bipolaron) is a metastable state because its minimum energy is higher than that of the two separate particles.

Discussion

We have proposed a model that describes the concept of autolocalization. The advantage of this model is that it allows us to study the basic concepts of autolocalization without using complicated and mainly numerical calculations.

We have shown that localized states can form in a regular dielectric or semiconductor with a sufficient (but not too high) density of current carriers. Only one (ground) level can be formed in a polarization potential well. If the dielectric constant is larger than 1.55, bipolaron states could exist as metastable states, because the binding energy of two separate current carriers is larger than the binding energy of the two carriers in one bipolaron well. Thus, it is obvious that there exists only one state in our model of a bipolaron polarization well. The localization radius of a bipolaron is considerably larger than that of a polaron.

Of course, reality is much more complicated than the model discussed here. Several excited levels of a polaron and bipolaron can exist, which also influence the properties of solids [3,4].

The simplified theory of polaron and bipolaron has been published in [8].

Suggested Problem

Consider a charged particle in a vacuum. Calculate and analyze the part of the electrostatic potential that acts on the particle.

Solution: It is well known in Quantum Electrodynamics that near the point charge in a vacuum, the electrostatic potentials has a form [9]

$$\varphi(r) = (e/r) - (2\alpha e/3\pi r)[1.41 - \log(\hbar/mcr)]. \qquad (4\text{-}27)$$

Here α is a constant, m is the (bare) mass of the particle [9], and c is the speed of light in vacuum. Equation (27) is valid for r smaller than $\hbar/mc$ [9]. The first term in Eq. (27) describes the part of the potential produced by the bare charged particle, while the second term describes the part of the potential, produced by the polarization of the vacuum.

Because there is no self-interaction in nature, the part of the electrostatic potential that acts on the particle is (see Eq. (27))

$$\varphi_a(r) = (2\alpha e/3\pi r)[\log(\hbar/mcr) - 1.41]. \qquad (4\text{-}28)$$

This function has a minimum at $x = x_e = \exp(0.41) = 1.51$, which corresponds to $r_e = 0.664(\hbar/mc)$. The minimum energy is $e\varphi_e = -3.01\,\alpha e^2 mc/3\pi\hbar$. At $\alpha = 1$ for an electron $e\varphi_e = -1191$ eV. It looks like that at such a deep energy minimum the wave function of the particle can be localized within the sphere of radius r_e. However the potential of the well, $\varphi_a(r)$, in the vicinity of the energy minimum is almost zero for $r > r_e$. So

39

the well is very narrow and the localization at $r \approx r_e$ is not possible as expected.

References

1. H Frohlich, H Pelzer, and S. Zienau, *Phil. Mag.*, **41**, N 314, p. 221, 1950.

2. SI Pekar, *Investigation on the Electronic Theory of Crystals*, Nauka, Moscow, 1951.

3. AS Alexandrov, and NF Mott, *Polarons and Bipolarons*, World Scientific, Singapore, 1995.

4. IG Austin, and NF Mott, *Adv. Phys.*, **50**, N 7, p. 757, 2001.

5. LD Landau, and EM Lifshitz, *Quantum Mechanics*, Pergamon, Oxford, 1984.

6. LD Landau, and EM Lifshitz, *Electrodynamics of Continuous Media*, Pergamon, Oxford, 1984.

7. C Kittel, *Introduction to Solid State Physics*, Wiley, New York, 1996.

8. Y Kornyushin, *AJP*, **71**, N 3, p. 229, 2003.

9. EM Lifshitz, and LP Pitaevskii, *Relativistic Quantum Theory, Part 2*, Pergamon, Oxford, 1986.

5. On Angular Momentum and Magnetic Moment in Many-Electron System

In this short section the author would like to comment a problem of statistical behaviour of many-electron system in external applied magnetic field [1,2]. It is stated in [1] that in Classical and Quantum Statistics application of external magnetic field does not lead to the changes in magnetic properties of a sample (this is a well-known Bohr and van Leeuwen theorem [1]). This means that it is stated in [1] that Classical and Quantum Statistics could not describe phenomenon of diamagnetism in many-electron system. It is stated also in [1] that many-electron system should conserve initial zero angular momentum after magnetic field is applied. Taking into account this conservation makes it possible to describe the diamagnetism phenomenon in many-electron system [1]. As is well-known [3,4], the angular momentum of an individual electron, $\mathbf{L} = m\mathbf{r} \times \mathbf{v}$, whereas the magnetic moment of it, $\boldsymbol{\mu} = -(ge/2c)\mathbf{r} \times \mathbf{v}$. Here m is the electron effective mass, $\mathbf{r}$ is electron coordinate vector, $\mathbf{v}$ is electron velocity, g is Landé g-factor, e is the elementary charge, and c is the speed of light in vacuum. It looks like electron magnetic moment is proportional to its angular momentum. It is so for an individual electron. It should be noted here that the individual electron angular momentum is proportional to its effective mass, whereas the individual electron magnetic moment does not depend on its effective mass at all.

It is well-known that many-electron system in a solid state can be described statistically, in particular, as many electrons, having different effective masses [3]. It is well-known that the effective masses can be of different values, and that the effective mass can be negative due to circumstances [3]. When there is a single

41

effective mass in a system, for such a system magnetic moment of a system is proportional to its angular momentum. From this follows that for such a system with zero angular momentum, its magnetic moment has to be zero also.

In real systems, as is well-known, in energy regions far away from minimum effective mass can be negative and of a comparatively very large value [3]. In metals, in particular, the majority of the delocalized electrons are in energy regions far from minimum. So it looks rather a common case for delocalized electrons in metals that they have total zero angular momentum (the sum of the individual angular momentums with different positive and negative effective masses), while the total magnetic moment of a sample (the sum of the addendums not depending on the effective masses) can be of essential value.

The specified here explanation of the coexistence of zero total angular momentum and non-zero total magnetic moment in many-electron system seems rather simple and relevant.

References

1. JH Van Fleck, *The Theory of Electric and Magnetic Susceptibilities*,
   Oxford University Press, Oxford, 1932.
2. Y Kornyushin, *Metallofiz. Noveishie Tekhnol.*, **34**, N 12, p. 1603, 2012.
3. C Kittel, *Quantum Theory of Solids*, Wiley, New York, 1963.
4. AI Ahiezer, VG Bar'yakhtar, and SV Peletminskii, *Spin Waves*, North Holland,
   Amsterdam, 1968.